Hearing Voices? Need Help?

Practical And Sometimes Humorous Guide To Dealing With Voices

978-1-312-21126-1
Lynx Taylor
5/14/2014

Prologue

    Are you or someone you know hearing voices? This book is here for you, to talk to you from your unique perspective with real world techniques tested and proven to work. This book is not here to say you are imagining things since maybe you are, maybe you are not. This book is written to consider all the various possibilities, and talk to you from where you are instead of trying to place you in a box, this book is written to think outside the box.

Chapter

Chapter 1. The Theories on Why People Hear Voices

There are all sorts of theories on why people hear voices. I'm going to include all the theories I have heard, not only the theories taught in psychology classes. The reason is people believe different things, and to lessen one reason over another bothers people. We are going to consider all the theories. You have probably heard some of the theories before, and we will break off into chapters where you can read whatever theory you believe to be the truth. So Why Do People Hear Voices?

1. Radio Waves and the Government (Our Government or Theirs)
2. Imagination
3. Telepathy
4. Brain chips or the cellphone theory
5. Angels
6. Demons
7. God
8. Memories
9. Brain Malfunction

In the chapters in this book each of these theories is considered, and ways of how to deal with voices in consideration of these theories is presented.

Many people will stick with number 2. in reasoning, and this is fine, but please be considerate of what other people think, and should your opinion ever change or you are dealing with people with different views than your own please consider reading the other chapters. This may help you talk to people from where they are, instead of trying to force them into your opinion too quickly or rudely. It is of great importance to listen to people, and not be so judgmental that people are driven away from possible help or techniques in dealing with their situations.

Chapter 2. People, Religious Beings, The Imagination, Or Aliens

Now let us discuss the various theories of where disembodied voices come from in reality. The first explanation is rather simple, yet complex.

1. People at various different places and times.
2. Religious Beings: God, angels, demons, and the devil
3. Imagination, Memories, Dreams
4. Space Aliens

The first two are outward sources, the third is an inward source for the voices. Please, remember I promised a humorous guide to surviving voices so I'm listing space aliens separately from people otherwise I would have lumped that topic in with just people. Now no matter where it is believed the voices are emanating from my first advice is in essence teach them Christianity. Jesus paid to take away everyone's sins so all who believe he died and lives again will at some time be in heaven. Why should you teach them Christianity?

1. If people are the source of voices, they tend to be more polite after learning Christianity **or** irritated beyond comprehension they cease to bother a person anymore after the person tries to teach them Christianity.

2. In the case of God and various religious entities they would be pleased or pissed you know so much about the Christian religion and it could either drive the voices off or explain why they are talking to you in the first place by reading all of the Bible. Learn to distinguish between the good voices of angels and the bad voices of demons.

3. In the case where the voices are from the imagination then there are various reasons why they would originate. One of these reasons is you are seeking to be loved or cared for in the deepest depths of your soul. God cares about you. In learning Christianity you may satisfy your inmost needs, and come to a time when you no longer need to hear the voices.

4. Space Aliens. All the reasons listed in Number 1. (politeness, etc.) And Number 2. (Are angels and demons space aliens? Read the bible to find out!) Present them with a part of our culture. Let them learn Christianity. This could really help you in negotiations or in forgiving closer than normal encounters.

No matter where your voices are coming from teaching them Christianity to begin with is a win, win, situation. Why Christianity and not other religions? Christianity has two qualities essential for a person to receive help: Caring and Forgiveness.

1. People need caring and forgiveness in their lives or they wouldn't be talking to you.

2. God would be pleased to know you understand Christianity so well and there is a way to cast out demons mentioned in the Bible in case it is demons

6

or the other guy, Satan. "Out in the name of God", fasting and prayer, etc.

3. When it is the imagination, memories, or dreams place them to rest in learning how to forgive yourself, and know God cares for you beyond all doubt.

4. Space Aliens, foreigners make mistakes all the time. Forgiveness is really important, and there is nothing wrong with teaching a foreigner to care about you, the earth, and well everything. Teaching Christianity is a good beginning provided no one is shooting at you at the time.

Now how to deal with voices truly depends on how a person believes they are being heard inside their heads, and who is behind the voices. When one believes oneself is behind the voices, rather than other people, then the techniques to dealing with the voices are going to be different. No matter what your theory is I will recommend music. Why music?

1. The majority of people like music. So compare what you like in music to what other people like in music. It is your life so don't let them bully you into listening to any music you don't like, but who knows, you might like some of the music other people recommend.

2. God created music, and it is definitely going to please him whether you are listening to Classical, Rock Music, Country, or Heavy Metal.

3. Stress is at times listed to be a cause of hearing voices. Music tends to relax the body and mind.

4. Space Aliens definitely could benefit from sharing in earth's cultural music. Music is said to be one of the universal languages. Communicate with music.

Now we have covered the four possibilities to who the voices are, and given two very effective methods of dealing with the voices. The question is who do you think is behind the voices? This is not a book where I'm going to tell you "you're wrong" because frankly you might be right. It might be your imagination, other people at different times and places, religious beings, or aliens. I am not here to judge you only to help you deal with the voices.

So what do you want to happen? Do you want to be able to think conversations without having huge battles? Do you want to live peacefully and quietly with your voices, or do you want them completely to stop?

Now if you want the voices to completely stop, I recommend you read this entire guide whether you believe it is only your imagination or have some different belief as to the origination of those voices. Why? Because one of the other chapters might hold the key to your relief, maybe, you thought it was one thing, but it is actually something else. Not to mention some chapters contain more humor than others, and if you're stressed about reading this book maybe those humorous chapters will mellow you out a little and give you some relaxation.

The two best ways of dealing with voices have already been listed but I will list more specific ways of dealing with things based on what you believe the source of the voices is in the following chapters. I list many of the ways to deal with voices in the end chapters, but in case I may have missed one or you wish to read in more detail about a method, I have made separate chapters for the various beliefs or theories regarding where voices originate from in this world.

With regard to humor please do not think I am laughing about any of the theories, but please understand people might and have laughed about the various theories of where voices come from, even the imagination theory has been laughed at in various cultures and time periods. So please stay with me while I take us through the various chapters. You may not agree with some of the theories, but other people might, and it is no bad thing to know how to advise other people concerning their own situations and/or beliefs.

Chapter 3. What if it is God? Or The Other Guy?

God usually accompanies his voice with a sign, a sign impossible for humans to accomplish. Do not be satisfied with a stop light turning green, a human or chance could do this. Read in the Bible about the various people who have heard from God and what signs he did for them to show the voice was God's: Hezekiah, Gideon, etc. I'm not going to tell you you're wrong since plenty of people have heard from God, but if the voice is asking you to do something evil or against the Ten Commandments it is the other guy and you should think "no". Who is the other guy? Satan. Satan can appear as an angel of light. The Bible tells us this. Satan is called the Anti-Christ since he is a deceiver and tries to pretend to be god. So you have to be careful when hearing a voice. Make sure it is God's voice and not Satan's. If he wants you to kill someone it would be a demon or Satan's voice. Do NOT do it! Think "no".

Learn Jesus paid to take away your sins by dying on the cross for you and lives again. With this belief you can request God cast out Satan by saying or thinking "Out in the name of God" or you can say prayers for protection. What to do if this does not work? Fasting and prayer is possible but only if you fast safely (no more than 40 days and it should be way less). Stay within the time common sense allows for, or consult a doctor on how much you can fast safely, etc.

Be Safe. Prayer needs to go with fasting and you need to be a Christian or this won't work. What if it does not work anyway? Consult other chapters in this book because someone is probably messing with you. Consider it might be a human at a different place or time, a brain malfunction, or there maybe other possibilities.

Chapter 4. Angels and Demons?

Joan of Arc, a great leader in France, claimed she heard angelic voices. She, also, was burned at the stake, unfortunately, so be careful who you tell about this, however, I'm not going to tell you you're wrong. Why? I have personally seen an angel with two white wings, black hair, and a white robe. I did not hear the angel speak, but God sent the angel in answer to a prayer I prayed to help particular people. I asked if I should pray for another angel to come, but the people said one angel was enough, and the answer to their cries for help. So I know angels are real and prayer works. This is why I recommend the Christian religion so strongly due to the fact my prayers on the particular day were answered. This is, also, the reason I do not make fun of people for their different beliefs.

Now if it is angels, great, but please be careful while driving and listening to angelic voices, but again if the voices are saying to do anything evil or against the Ten Commandments consider they might be deceptive demons instead, trying to hurt you. There are stories of possessed trying to thrown themselves into the fire, etc. I do not want you to be burned so I'm going to teach you to request God to cast out a demon or demons. Please, read the Bible if you want more instruction on how to cast out.

First you have to be a Christian. What does this mean? This means you believe Jesus died on the cross to take away your sins (wrongdoing) and lives again so one day you will be in heaven. Once you believe this you can say or think "Out in the name of God" and God will cast out demons for you. Now a couple of tips: One, it is always God who casts out the demons, not you, so remember this. Two, unbelievers once tried to cast out demons, and the person possessed with the demon beat the crap out of the unbeliever trying to cast out the demon, so make sure you believe before trying this.

What if it does work but then does not work? Here is the problem: God can cast out a demon but if the person is not a Christian the demon will be back with more demons to possess the person. Look "possessed" up in the index of the Bible and read all about it if you do not believe me. This has happened, so what to do? In the moments after saying "Out in the name of God" teach the person the essentials of Christianity: Jesus paid for everyone's sins, etc. You may have only minutes before the demons try to retake the person, and you have to say "Out in the name of God" again. How do I know this beyond what is written in the Bible. There was a woman who was acting weird and almost attacked me. I said "Out in the name of God" and she stopped, but then a few minutes later tried it again. I said "Out in the name of God" and then explained the gospel to her. It did not happen again.

What to do if this does not work? Two possibilities and one is to try safely-fasting, and prayer, or read one of the other chapters in this book. Sometimes a person thinks it is one thing and it is something else. Chapter 14 has a list of methods I hope will work for you.

Chapter 5. Radio Waves, Telepathy, and Brain Chips?

Beginning with Radio Waves and the signal areas of the brain is it possible? Is Wi-Fi possible? Years ago having a computer on the internet without a chord was considered impossible. Now we have Wi-Fi so who am I to tell you it is possible or impossible. However, if you believe this is the case then I think you should study how to block sound waves safely should you wish to be rid of the voices.

There was a girl who carried with her an item of copper or would walk in front of an art piece wall hanging made of copper whenever she heard voices. Why copper? She said the copper received the signal instead of her. Sometimes she had to block two areas where the signal was coming in from but this actually worked for her. Does tin foil work? I have never heard of it working. Do other metals work? Possibly, I've heard people having luck with certain metal alloys, but to understand it farther, study what blocks radios and sound waves in general, and there are natural areas where sound waves are blocked. So go buy a text book on communications concerning technology and best of luck to you.

Telepathy is it possible? There are studies made on the brain saying we only use a certain percentage of it. What if telepathy is using a higher percentage of our brain? What if certain people are simply unblocked in the area linked to telepathy? I've seen a great deal of weird things in this world so I'm not going to say you're wrong because it is possible. What I'm going to write here is, please do not think you can read everyone's mind. Talk to people out loud about what they're thinking and never assume you know. Why? Even if you're one of the lucky ones whose brain is unblocked this does not mean other people's brains are unblocked or active in the area of telepathy. So do not assume you can read their minds. Chances are you cannot since they are only using a certain percentage of their brain. So is this theory correct? I do not know. I personally do not have the gift of telepathy, but there are people who have sworn they do have the gift of telepathy and if you are taking this view of hearing voices then I'm glad you take the view of hearing voices to be a gift instead of a detriment, and I want to encourage you in believing this ability to be a gift and not a detriment. Given this most psychologists and psychiatrists are going to disagree with you but they have their opinion, and you have yours. Use your gift wisely.

Brain Chips can be a gift too. I'm not advising you to go out and get chipped even if that were possible to do on a normal basis, but what I'm saying is please do not view it negatively and try to surgically remove the chip or bang you head against the wall to try to break it. These ideas are both bad and will land you in a real hospital, dead, and/or in a mental hospital. Not to mention both are painful ideas. Since you probably are not a surgeon and even if you are operating on yourself is probably a bad idea please leave the chip in since it can be a benefit not a detriment.

People giving you orders driving you crazy? Give them beneficial orders instead. The people listening probably do not know who is who, so giving them orders instead, often causes a little confusion, and can give you a breather. Not to mention if you view this as a brain malfunction or your imagination then you should be the one giving yourself orders, not unknown voices. If they are telling you to do anything evil: Think "no".

You're in charge of your life, not the voices. So what should you do if you've been given the gift of a brain chip? Use it wisely. I do not care if you think it is agents from a foreign country or space aliens who gave you the chip this is an opportunity to teach them Christianity. The people behind the voices will never see it coming. Just think, you could be reaching people as far away as China or North Korea. They gave you the chip so it is their fault if the people in their regions become Christians overnight. This being said it is difficult to teach Christianity but very worth it.

Chapter 6. Memories and Dreams

Are memories and dreams how you are hearing voices?
You are likely very normal then. Everyone remembers
people speaking in their past and people normally do
hear voices in their dreams. Now in severe cases it is
probably not schizophrenia but P.T.S.D. or Post
Traumatic Stress Disorder. I'm not a doctor so I'm not
going to diagnose you, and do not take this book as a
diagnosis, it is not. But if you think you have Post
Traumatic Stress Disorder and are working with firearms
please be brave enough to seek help immediately if you
are afraid you are going to shoot someone else by
accident or in rage. Please, be as calm as possible.
Seeking professional help is not for everyone, but for
some it can benefit everyone around you.

What this book cannot do is be a good listener for
you, and someone who is trusted and a good listener is
exactly what a person who has P.T.S.D. needs.
Professional help means they might assign you to take
pills or in extreme cases lock you up in a mental
hospital if you are deemed to be a threat to yourself
or society. Forgiveness is needed in this instance
since you have to ask yourself what you would do if you
were unsure of someone who has firearms?

Not working with firearms, but still feel you might
have P.T.S.D. then it is your decision whether you seek
professional help or not, but if the voices in the

dreams are leading you to do anything evil then: think "no" and check out some of the other techniques in the other chapters to help you with this.

Spirits in your dreams or memories, ghosts haunting you? Please, read the chapter on angels and demons if this is the case since angels and demons are often referred to as spirits.

Chapter 7. Imagination

Let us begin this chapter with the idea if it is your imagination you should be able to shut the voices off and on the way you would opening or closing a book. If this is not the case I would ask you to consider reading one or all of the other chapters because I hesitate to believe it is your imagination if you are unable to shut the voices on or off.

So you can shut the voices on or off. Congratulations you have imaginary friends. So what we need to work on is gaining some real friends to substitute for those imaginary friends. Socialization is what you need. So think about what you can do to go out and talk to real people. The best way to find a friend is to be a friend.

What about learning a new hobby, taking up an internship or volunteer work. Learning something new will place you in contact with people you can be a friend to, and who perhaps may become your friends, too. Learning a new skill, hobby, type of work beneficial to society and enjoyable without being in trouble with law enforcement, for instance please ask the owner of the building what type of drawing or words they may want on it should there be any urges to spray paint words or landscapes on the sides of buildings, walls, etc. not belonging to the artist. Who knows the

owner may even donate some paint, possibly, (never
expect it), pay the artist for their artwork provided
they like the words or pictures be careful not to break
the law concerning those pictures.

Chapter 8. What if it is the FBI or the CIA? (How not to annoy agents.)

   Be smart about this all. Let us begin with saying you are under no obligation to help these people. You can say "no", and if they are from North America they should respect your reply. In real life there was a young man in a foreign country who was asked by the CIA to use his camera to spy on what was going on in the country he was traveling through. He took the smart route, and said "No".

 Why is this the smart route, and how could he have possibly said "No"? It is all fun and games, right? It is not fun and games. This young man who did not hear voices, but was asked point blank to use his equipment while on vacation to spy on someone said "no", and this is a true story.

   This is real life, and being involved with helping the FBI or CIA, or any agent from any of the various foreign governments can have real consequences. You could be drugged, beaten up, bones broken, arrested, or killed. Your entire family could be in danger because you tried to help someone. Not to mention they will throw you in a mental hospital should you admit to be hearing voices where they will prick you with needles for examinations, drug the crap out of you, and disbelieve you. Why?

Some would say the left hand of the government does not know what the right hand of the government is doing. In other words should the technology exist for mind to mind communication or the possibilities of telepathy being explored prove to be real then it is so incredibly classified, street cops and ordinary agents would not even know about it. Therefore if you are hearing voices, and think they are working for a government be careful.

First of all how do you know it is your government? Foreign agents can speak or think in English. Secondly, even if it is your government you have the right to say "no, I'm not going to help you with this," and think out your reasons. "I don't want to go to a mental hospital" or "I don't wish to get involved because I have no guarantees my family would be safe."

Do not ever hit your head against the wall. If you have hit your head against the wall you know it hurts. All the things I've described in the previous paragraphs including broken bones hurt too. So be smart about it, and think to them "no, I'm not going to help you. I'm not an agent, I'm not trained for this, and I'm not sworn in to help my country."

Again this is not a book where I am going to tell you you're wrong since maybe you are not. So after considering the consequences and everything in the first few paragraphs of this chapter you decide to help anyway. Maybe, all this does not scare you and you are not afraid of mental hospitals, etc. There are a few things to consider before trying to help the FBI, CIA, or unnamed agents. First of all they will never acknowledge you in public, not because they do not like you, but because it is so classified they cannot. So do not expect them to get you out of jams or mental hospitals. I'm not saying they can't or won't, but respect the fact of life being they are there to do their jobs, not go off on side missions not sanctioned by the government to get you out of the mental hospital.

Simply put, there is no special code I'm aware of wherein they tell you, "you tell the mental hospital, and they just let you go." If there were I think you should ask for the code to begin with in your negotiations concerning helping them. Secondly, how do you know you're helping our government agents? How do you know they do not work for North Korea or Russia? Honestly, the United States of America is not number one in technology. What credentials are they presenting to you to show you they are on your side or your country's?

Do not settle for a badge number. Anyone could look at a police badge and rattle off a number. Really, think about what you're being asked to do and would it benefit your country or not? Don't do anything illegal. If they are really agents from your country they are not going to ask you to do anything illegal especially since you do not have the training they have.

1. Good or evil. If it sounds evil, do not do it. Think "no."

2. Is what they are asking you to do going to attract attention to you? You do not want to end up in a mental hospital so your safest bet is to think "Sorry, guys. This would really be unsafe for me to do so I'm not going to do it." If they are the good guys I think they will understand and you might even earn their respect.

3. Is it a party line? Are they actually talking to you or is it another agent they are trying to contact? In the old days phones worked off what was known to be a party line where everyone could hear each other's phone calls. If this sounds like the case then do not assume they are talking to you.

When it comes to calling 911 over voices in your head don't do it, unless you can physically see the danger. Why? You could be hearing voices as far away as Afghanistan or there might be a fire they want you to respond to in Switzerland! Your emergency responders are just not good enough to home in on a voice and know where it is coming from, at what time it is coming from, and bring enough aid to the scene.

Think or ask the questions:

1. Where?
2. When?
3. Who?
4. What or How?
5. And Why?

Real agents are going to respect you for wanting to know even if they can't tell you everything. The "who" part is likely to be ignored but it is still important due to being able to identify who are the bad guys, at least, and who the agents want you to trust. On the issue of trust, don't trust anyone who says they are an agent unless they have just saved your life, even then I'd still be wary of them. Why? People lie.

Agents know how to lie. This means whatever they are telling you in your head may or may not be the truth.

So be careful in trying to help them so that you don't fall into a trap. Such as, maybe foreign agents wanted to steal records from the police station and have you call in a false emergency so they can clear out the place in order to steal those records or make a robbery down the street. This is just an example, obviously in real life there should be enough police to defend the police station and check out a 911 alert, but what I am trying to explain is sometimes it is a false alert the voices in your head want you to call in and this could possibly lead to you being arrested or taken to a mental hospital. Be aware of this. Another possibility is there was a real crisis last year or a year ahead of when they want you to make the call.

This is why the Where, When, and What are so important. Even given the W's I would not report it unless I actually see flames coming up from that location. Why? Because again people lie, and consider, agents watching a house or public area, sometimes they get bored and want to play games with people. Cruel but it is simply human nature. What if they are so bored they just randomly start giving you commands? You don't want to be the butt end of one of their jokes so think "no".

What if they in a public place want you to turn your cup upside down or otherwise indicate who you are? Think "no" because by turning your cup upside down you've just exposed yourself to a possible assassination attempt or told them "I'm an idiot who will do whatever you say". You're not a robot. Make the people behind the voices respect you, and be smart. Why would they ask you to turn the cup upside down? Are they telling the truth or are they lying? Think about it. Do not just do what people tell you to do or you will end up banging your head against the wall and your head will hurt!

What if you have proof: you have confirmation and agents showed you credentials in real life. Then they vanished as they tend to do since they are agents! Do not expect them to back you in any explanations whatsoever. Do not talk to people about it. Do not expect people to believe you.

Even if you helped them in a super secret mission they are not going to wave in a friendly way back at you (there are a few exceptions but generally those are agents with a sense of humor). You are likely never to hear from them again and everything you helped with concerning head voices, etc. is highly classified. So do not talk about it. They are probably not going to talk about it. So you should not either. Why? Because people are put in mental hospitals over these types of things to discredit what they say about top secret things or people simply do not believe you. So be smart. Do not talk about it.

Try not to do anything too extensive concerning helping with situations. Attracting attention to yourself is bad, but if all they really want you to do is help take a cat out of a tree for a neighbor this would be a good deed and I see nothing wrong with doing this action provided you have a ladder and are not afraid of heights, though there is the risk of falling and having a broken limb. Consider the risks when you are asked to do something. Is it a reasonable thing to do these voices are asking you to do, or is it unreasonable. If unreasonable think "no".

Be smart about what actions you take or do not take. If all the voices ask you to do is help an old lady across the street this means they wish they could help but are busy with other things, and could you just help the old lady across the street? Simple actions not causing any damage are alright to do for people, but beware of more complex actions requested from disembodied voices. Simply think "I'm not an agent. I cannot help you hack into the Pentagon, besides how do I know you're not a foreign agent who is trying to get this information?" for example. You're not trained to do these things and even if you are, would doing this actually benefit you or your country?

Think about it first before taking any action any voices urge you to do. Be smart.

What about physical threats or the thought "They're after me?" what do you do?

First do not act paranoid! Why? This is the reason people will come and lock you up in a mental hospital where all you can do is say prayers to help you or your family. So practice acting cool and calm no matter what the voices tell you.

Please, practice remaining calm if you do not know how at first. Remember an actual agent would remain calm. Paranoia is a problem. The last thing you need to do is give people a reason to lock you up. This being said what should you do?

If you have to do something weird such as look under the car seat to check for a bomb have a logical explanation having nothing to do with the idea there might be a bomb under the car seat. Such as "Oh, I thought I'd check under the seat for my wallet" or you could say "loose change". Something reasonable not drawing too much attention to yourself. Repeated paranoia or weirdness will be noticed so even if you have logical replies should your behavior continue in paranoia then people will lock you up based on your behavior so try to keep the worries over car bombs, etc. to a minimum. Notice if the voices lie once to you about something, chances are they are probably lying again. For instance "No car bomb" detected under the seat? The voices lied to you. Do not ever take their word for it again, and you do not have to check every little thing since obviously someone was trying to make you paranoid so they could lock you up in a mental hospital. Yes, it happens. People might actively be trying to drive you insane. I would not assume they are agents if I were you but it is still possible.

We have probably all heard stories where thieves wanted the mansion property so they installed hidden speakers, and projectors to make it seem the mansion was haunted. Why? To drive off the owners so they could buy the land at a lower price. Assuming this might be the case there is one particular trick that tends to work on people who are actively trying to drive a person insane by using head voices. This is the chant. Do not say it out loud, but find a quiet, safe place where you can think for an hour or two and then think a chant to stop the ceaseless voices. Why? Because this irritates the crap out of people trying to drive you insane, and they usually give up and leave you alone after a half an hour or less of hearing the chant "Def Leopard go listen to their cds", "Come out in the name of Jesus", or "Your sins are paid for by Christ, pray to him, and leave me alone". It can be any type of chant. Make one up, provided it is something you can think for a half hour (it usually only takes fifteen minutes or less), but it has to be something you can handle thinking for a long time. They may try to chant back but this is your turn to tell them to go do something, remember this, it is your turn to drive them insane. The best chants are usually something helpful for you to listen to or even for them to listen to. Such as "Keep your thoughts on what is lovely and beautiful" or "Go listen to Heavy Metal" something simple and effective. When asked by people concerning what you are doing simply say "I'm trying to sleep", "meditating", or if you have headphones on "I'm listening to music and singing along".

So what if it is a real threat and they are not just trying to drive you insane? What if people actually start towards you threateningly and you don't know what to do. This crosses the line between what is believable to cops and what is not. Hearing voices is not believable to cops so don't ever try to tell them about it and expect any type of help beyond the cops trying to lock you up in a mental hospital. So what do you say when people are coming threateningly towards you?

Go stand by a policeman if there is one available in the U.S.A. (in foreign countries obviously what I advise may be different concerning your safety based off where you are). Do not tell the policeman you are hearing voices. Do tell him there were people moving towards me in a threatening manner (only if there were people there moving towards you in a threatening manner), and I just thought I should maybe stand by you for awhile in case there is trouble. Be able to describe the people. If there were threats in your head by the voices simply say you overheard someone saying whatever the threat is but you weren't sure they were serious. He will probably ask you where you overheard the threat and by who. Be honest, but do not tell the policeman everything. Say I overheard the threat while standing there (point to location if within sight), and I am not sure who made the threat because my back was turned towards them or something similar. Be honest, brief, and do not tell them you are hearing voices in your head if there is a real concern for your safety because the minute he hears you say "voices in head" is the minute he either turns away or makes arrangements to have you assessed and locked up in a mental hospital. I do not care how nice the police seem they are not going to give you any help regarding disembodied voices except to have you examined and locked up into a mental hospital.

So do not mention the head voices to them only mention what you can see is happening for sure. Be very clear on the details and do not lie, just do not tell them about the voices. If you really have to repeat something one of the voices said in your head say "I overheard, but I'm not sure if what was said was true" or "I heard it on the cellphone" if you have made any previous phone calls. The latter explanation is not recommended in this day in age since they can trace your phone calls and find out exactly what was said in some instances so use it only if you are completely desperate.

Logical explanations are always needed whenever there is a police matter so be brief, logical, and calm. Do not lie. It is really hard to remember what you said after the fifth police officer questions you. Try not to report anything unless there really is a seen physical threat to your person. Voices tend to play games with people and say a threat is from one person when the person indicated had no intention of threatening you and probably does not even know you.

Remember you do not know who the person behind the voice is. So do not assume who it is even given the Who, When, Where, What questions have all been answered because, remember, people lie. So be really careful when interacting where voices are concerned and just thinking something should be enough, you do not have to say anything out loud to contact the people behind your voices.

When it is excessive voices talking in your head I recommend listening to music, and/or teaching the agents Christianity. Why? Because it is less likely they will ever shoot you if they become Christians, and due to the type of work, they deal with a great deal of scumbags, so it is helpful for them to be able to forgive themselves for some of the actions they have had to take in the past.

Chapter 9. Brain Malfunction

This is the saddest possibility, and the one most often sited to be the truth by psychologists and psychiatrists. To begin with I want to assure you we humans heal. Given time, you should be able to regain what you think you have lost. The mind has an incredible capacity to heal, given rest.

The first three things I'm going to recommend are sleep, eating fish, and prayer. Why? Most often the brain needs time to recover and resting or sleeping play huge roles in this recovery. Why fish? Fish is said by nutritionists to help the brain. Prayer? God is the greatest healer there is and if anyone can help bring your mind back to the way it was before the brain malfunction, it is God.

What else can you do to help your brain function normally again?

1. Practice doing things normally the way you would do them were there no voices in your head.

2. Relax. Rest and sleep are vital for you in order to regain the normal functions in your mind.

3. Focus on prayers. This helps a person focus the mind on where you are at now, instead of wishing to be somewhere else or stressing about what you cannot yet do again.

4. Understand, God loves you no matter what condition you are in and you do not need to hear voices to know God loves you or cares for you deeply.

5. Listen to music you like. This may help bring the brain back to more normal functions.

6. Chant and meditate to focus the mind for a few minutes, when you have the time to do so, to relieve stress. You can think the chant instead of saying the chant out loud and the chant can be anything from Semper Paratus (always prepared, the motto of the U.S. Coast Guard) to The Lotus Blossom Blooms in Spring, to a Christian prayer.

7. Healthy Lifestyle (certain vitamins are recommended to treat some conditions such as Schizophrenia)

Now I am not going to spend a great deal of time on Brain Malfunction advice to return to normal and this is the reason: many of the techniques in the other chapters work well even if you believe it is a brain malfunction so I recommend you read them simply to help strengthen your comprehension and for the sake of laughter. Laughter is sometimes worth more than sleep in healing your mind.

I am not encouraging you to laugh at the other possibilities of where voices come from, but I realize some people will anyway, and I did write some humor into the chapters. If it does not make you laugh then pick up the comics and read them! Laughter can help heal your brain and I definitely recommend it along with a healthy lifestyle, and receiving the proper

amount of sleep or more. Brains need time to heal so do
not be frustrated; allow yourself time for healing.

Chapter 10. Space Aliens? Are they real?

This was included in the book due to the large percentage of people who truly believe it is space aliens talking to them. The first question I have is if this is true: How did they learn English? Please understand I am taking your inquiry seriously the way I take all the possibilities in this book.

So the space aliens speak your language. This is a given since you would not be able to understand your voices if they did not. Let's examine all the possibilities then.

1. Space Aliens exist
2. It is other people playing an elaborate hoax on you
3. It is angels and demons (are they space aliens?)
4. It is your imagination?

1. Beginning with the "space aliens exist" theory. Again the first thing I would do is teach them Christianity. Why? There are bound to be misunderstandings in negotiations or conversations so, teaching them to forgive you, is not a bad plan and you being able to forgive them is necessary. Now you would have to assume they are clever, (after all they already know your language), so this means they know how to tell the truth and how to lie.

I would highly recommend you read the other chapters particularly Chapter 8. Why, because you're going to be dealing with a great deal more than aliens if you believe they are real. Understand, I am a skeptic when it comes to space aliens so I write this with the same

way I would write about Chapter 8. Beginning with the
W's so highly useful in Journalism.

1. Who?
2. Where?
3. When?
4. Why?
5. What or How

Always what I begin with when dealing with people,
and you are going to want to be non-prejudicial towards
them, so let us call them "people". When dealing with
people you want to be reasonable. If they ask you to
climb a high tower tell them this is unreasonable, and
refuse their request. Why? Because you're not a moron
or their slave, and if the high tower is on private
property you are going to get in trouble for being
there and possibly locked up in a mental hospital.

Avoid vandalism. Trust me if the aliens are so smart
they can understand your language then they can cause
the vandalism all on their own without your help! I
would follow what I said in Chapter 8. You do not have
to help these people. Tell them "no". Gain their
respect by showing them you are your own person, not a
pawn. You need to understand, most people will not
believe you and if you talk constantly about aliens
there is a great possibility you're going to end up
locked in a mental hospital at some point in your life.

Did you see a flying saucer? Telling the sheriff is
simply not going to help you. If he did not see it,
(unless he is into believing about aliens, which I
highly doubt) do not tell him. No matter how nice he
seems, he is not going to believe you.

Do flying saucers exist? Yes, I believe they do exist. Perhaps not on the same scale as some who believe in UFO's, but I am not about to ignore what I did see with my own eyes. I have witnessed a saucer hovercraft at a science exhibit at my school a long time ago so if they can make a hovercraft out of vacuum cleaners in the shape of a saucer then God knows what the various governments around the world have been up to. Not to mention the various shapes of the landing pods we use on our spacecraft which tend to look the way saucers do at certain angles.

Been experimented on by aliens? My sympathies, begin to hear voices after one of those incidents? Please, read chapter 5. Messed up really bad? Forgive them and forgive yourself. When encountering higher technology there is little one can do except say prayers.

Do I believe in higher technology? Do you believe in cellphones? I grew up at a time when there was barely TV. Think of this in comparison where now we have the internet and cellphones. There is higher technology out there, and I'm not going to dig myself into a little dark hole and pretend there is not, nor should you. Concerning talking about your experiences, do so at your own risk. If mental hospitals appeal to you and their free medical care is needed then I'm not going to tell you not to go there, but a word of caution: No matter how nice they are they are probably not going to ever believe what you say. I'm not trying to be rude. I'm merely telling you the cold harsh reality.

So where can you turn to? There are many people who believe in aliens and UFO's on the internet and people who write books about their experiences, etc. They are out there and you can find them and talk freely with them concerning your experiences. In fact, some might even help you write your own book!

2. Perhaps it is other people playing an elaborate hoax on you. People are sometimes cruel or prankish and full of fun. They may well rent costumes (and some of the costumes now-a-days are realistic looking: example, if you've ever watched the wraith on Atlantis Stargate), hide microphones, and go all out to convince you there are aliens, but it is just other humans.

   Now I realize this may be a disappointing possibility for you but let us look at the facts. The aliens know your language. How? They could have been other humans all along just joking with you or being cruel. Usually it is pranks so please do not take it personally. Rarely have I seen them being cruel about it. What to do? Be skeptical. Be reasonable and logical. How do you think the voices are coming in? Short range radio waves? Brain chip like a cellphone, God knows those things are sure small enough to fit inside the head? Telepathy? All covered in a previous chapter. Again if you think it is a brain chip do not attempt to take it out. I'm sure if there really is one, it will probably set off the metal detector at the airport. So study science in communications. There has to be a logical explanation.

Number 3. It could be angels and demons, please check out the chapter on them. Now some angels are described to have six wings or multiple eyes. Is this a cool appearance for an alien or what? They are called Seraphim and Cherubim. Angels and demons are real beings according to the Bible so do they come from earth or another planet in the solar system? You can read all about them in The Bible.

Number 4. You think it is your imagination but you're
not sure. To begin with if you're hearing weird noises
and the noises are originating from your property:
(Please, don't trespass on other people's property.)
Provided it is not on the roof or in a hazardous place,
go check to see what the noises really are. It might
just be squirrels building a nest. Please, do not go by
hazardous machinery in checking it out, and avoid
climbing tall places. Studying radio waves in a text
book is a good idea. Again if the voices are telling
you to do anything bad. Think "no". Generally if you're
trying to substitute little green aliens for friends
read the chapter on Imagination and if you're imagining
big bad aliens then imagine something different such as
good aliens. Of course, if you've explored where the
noises are coming from and find someone beating on
someone else and this is where the screaming is from,
(if you can visually see this happening), go ahead and
call 911, but if you cannot see it happening but there
is screaming anyway, then think to the person screaming
to call 911. That is right-instruct the aliens on how
to call 911, but never do it yourself unless you can
see there is a problem.

Chapter 11. Avoiding Mental Hospitals (Be Safe)

The key to avoiding mental hospitals is good behavior. Primarily, unless you've been diagnosed with a mental illness in the United States of America they cannot place you in a mental hospital unless they can prove you are a danger to yourself or society. This may change or may be different in your state, but most of the U.S.A. has this requirement.

This being said, keeping cool and calm is a must, and avoid mentioning you hear voices unless you already are seeking professional help. Professional help means the professionals will likely place you on pills, schedule sessions with you, and in extreme cases lock you up in a mental hospital or suggest you should check yourself in willingly. I'm not going to tell you, you are wrong if you are seeking professional help, but I'm going to warn you, professional help is not for everyone. There are people who are helped by it. Then there are people who wished they had never seen a pill before. Much of it depends on if you can bond with the people trying to help you in trust or not. Most of my friends preferred talking to me instead of a psychologist or psychiatrist, primarily one of the reasons I'm writing this book.

So this is a chapter on avoiding mental hospitals. Be safe in this, in other words, if you have suicidal tendencies I'm going to recommend getting professional help, and understanding God cares for you very deeply no matter what. There will always be a better day, so keep on living. The reason I recommend professional help in this one case is this book cannot be there for you in the form of a good listener, and you need someone who can listen to your problems.

One technique in avoiding mental hospitals is singing or rapping. People may not understand you hearing voices, but what they might understand is you singing along to headphones or making up lyrics to your own tunes. This can be a huge benefit to keeping out of a mental institution by dealing with your voices in this way.

So the worst has happened and you are already in a mental hospital?

Again good behavior is the key to being released. Fighting them means you get more drugged out, and they place you in more restrictive environments. So practice staying cool and calm. This is a good time for prayer, and using any of the other techniques I've described in Chapter 14. Test them out. See if they work.

Chapter 12. What To Do When The Voices Are Screaming At You

Learn the phrase "Please, excuse me I need to concentrate." Or "Give me a few minutes; there is a problem I need to think through." Some people simply say "I have a headache." But I would rather a person did not lie about this. You do not have to tell them you're hearing voices screaming at you, in fact I would not at all say this, but only say you have a headache if you have a headache. These phrases give you a chance to walk away, and be somewhere quieter or alone.

Now depending on where you think the voices are coming from there are various techniques to deal with them.

1. If you think it is other people (if you think it is space aliens I'm including my advice for this, too, in with number 1. After all they are people too!)
2. If you think it is your imagination or a brain malfunction

Now if the case is number 1. Let's go back to the W's of When, Where, Why, and What or how. Ask them these questions in thinking them, not out loud, just in thought. What is wrong? When, where, why? Do not necessarily believe what they say, but take into consideration they might be telling the truth so if it is an emergency you can instruct them how to call 911. You never call 911 if all you have is the voices word for it, but you can think of how to, in order to help them in their crisis.

So what if the voices claim they are coming from Afghanistan and there is no 911? Think through what you would do in their situation with the most reliable advice you can think of, such as, if there is a building on fire how to safely exit the building, etc. Now if this does not calm down the screaming I recommend praying for the people screaming. If this still does not work there is the think chant. Thinking a chant, but not saying it out loud such as "It will be alright" or "Jesus paid for your sins, and you'll make it into heaven someday, wait for it" or whatever you can think up; a phrase you can think for awhile. If this does not work, I recommend listening to music or watching a movie.

What if none of this is working? Another possibility is to try and go to an area where radios and cellphones do not work. This is particularly used if you believe it is radio waves or a chip. Do not try to remove the chip, and do not hit your head against the wall trying to break the chip. These are harmful actions to you, and I've never heard of them helping, but there are ways to block sound waves. I suggest studying those methods and knowing places where the sound waves are naturally blocked. You can go to these places, please take a friend if it is a remote area, but it can help stop the screaming.

So on to number 2. So it is your imagination or a brain malfunction. If it is your imagination, imagine things going differently and people being safe to stop the screaming. You should be able to turn it on and off if it is your imagination. Now if it is a brain malfunction first of all you do not have to do what the screaming voices say. Simply tell them "no" if it is something evil the voices are screaming at you. This being said there are a couple of things you can do. First of all read the number 1. techniques because they all apply here but for a very different reason.

   Screaming indicates for some reason in your life you are fearful of some situation deep in your mind, underneath it all. So working through the situation the screaming is originating from could calm you down. Say the voices are screaming because of being burned. Imagine applying first aid to the burn with a white gauzy pad or salve if it is a minor burn. This is simply to calm you down. So what if you cannot understand the reason for the screaming?

 Think reassuring thoughts. Pray to Christ if you are a Christian. If you're not a Christian please read the chapter on angels and demons even if you don't believe in it. Sometimes, "out in the name of God," actually works. So what if it is not a demon? It likely is not but who wants to take the chance? Let us continue with the reassuring thoughts into the think chant. Think of a phrase you can repeat over and over in your mind, not out loud (since saying it out loud might alarm people around you). It should be something fairly simple such as "Please, stop screaming, it will be alright." Or "Jesus cares for you." Or "I'm going to listen to heavy metal music if this does not stop." Something very simple, but that you believe in thinking, and if this does not work I highly recommend listening to music you like or watching a movie. Voices tend to cease when watching a movie possibly because the mind is set with a different task in taking in all the visual projection and sound. So again I will list the things I recommend should the voices be screaming in a neat little list:

1. Learn to say "Please, excuse me, I need to concentrate right now" or "Give me a few minutes I have a problem to work through."
2. Think helpful or reassuring thoughts through the situation the voices say caused the screaming.
3. Prayer
4. Think Chant
5. Music
6. "Out in the name of God"
7. Watch a movie you like
8. Block the signal or go to a place where there is no signal

Chapter 13. How Not To Anger People Trying To Help You

People do not always know what to do when they discover a person is hearing voices. You are going to have to forgive them. Think-what would you do if someone told you he or she was hearing voices?

Again good behavior is key in trying to get along with people who are trying to help. Please be polite and understand they are not always going to do the right thing or have the right suggestion. People are flawed, not perfect, so do not expect them to be.

What about if people are talking about you out loud or in your head? Have people in the past? Calm the panic button down even when people do not have the best view of you. Perspectives sometimes change. Talking with people in a beneficial way when in a calm and polite mood can prevent a great deal of damage. Instead of thinking of the bad things a person has done, think of all the good the person has done. I'm not telling you to ignore reality. You still have to understand how the world looks to other people through trying to understand their actions from their perspective, in order to help interact in a more beneficial way. I don't recommend this to everyone concerning family and friends but if you were in their past or current situations what would you have done? Sometimes a little forgiveness is needed.

So the person gave you this book in order to try and help out? I hope the techniques in chapter 14 are there for you, and this book really does benefit you. Whether it does or does not, a simple "Thank you" to the person who is trying to help you can go a long way.

Chapter 14. General Techniques To Survive Or Stop Voices

1. Learn to say "Please, excuse me, I need to concentrate right now" or "Give me a few minutes I have a problem to work through."

2. Think helpful advice through the situation the voices say is causing the screaming.

3. Prayer

4. Think "no" when voices are telling you to do something evil

5. Music

6. "Come out in the name of God" said or thought by a Christian

7. Watch a movie you like

8. Block signals (study sound waves and how to block them from where you are without hurting yourself or others)

9. Go to a place where radio signals are blocked, take a friend if it is a remote or scary area

10. Only recommended if the voices are dreams. Find a good listener who you trust to talk out your problems with.

11. Find friends who speak to you out loud in person

12. Imagine a better outcome for when voices are screaming at you if you think it is a brain malfunction or your imagination

13. De-stress if you believe it is a brain malfunction or your imagination by ordering pizza, tea, coffee, or your favorite flavored non-alcoholic drink.

14. Play a video game. This helps de-stress and brings the focus back around to the video game instead of on voices. This is important for those who believe it is a brain malfunction or the imagination and is similar to watching a movie only more interactive.

15. Particularly for those who believe they are suffering a brain malfunction, or it is the imagination try painting, sketching, or drawing to take the mind off voices and center concentration on the work at hand. This is a good relaxing way to think of something else.

16. Only for those who believe it is a brain malfunction I suggest laughter to help take away harmful stress, (this does not mean laugh at nothing or what the voices are saying, in front of people, in fact, please never do this). What I do mean is read the comics or a humorous book to take the edge off a possibly stressful lifestyle.

17. Sleep. For those who believe it is a brain malfunction and even those who do not, please make sure you are receiving the proper amount of rest to be capable of dealing with voices, etc.

18. Singing

19. Think Chant

20. Teach the voices or yourself Christianity. (God cares about you)

These techniques are listed in other chapters, particularly listening to music. If there are any questions about a technique please read the other chapters describing the use of the technique and why it is listed. What I wanted to do for this chapter is list them all, and I hope I've covered all the techniques in the book. Please, forgive me if I have forgotten any, and I sincerely hope these techniques give you a good solid base to cope with hearing voices.

Chapter 15. No Fear

Having eternal life in Christianity goes a great distance towards having no fear. Jesus paid for your sins so you are going to heaven someday. This means God is with you no matter what. God is going to be there for you when you hear voices, if you have to go to a mental hospital, or even in the worst of times. Prayer instead of fears can help in the worst of times, and you will always have a Savior looking out for you.

Still don't believe? Then be aware of your surroundings and try the think chant to calm yourself down in times of severe stress or fear. I'm not going to tell you what to believe or condemn your thoughts, what I'm going to do is present my own perspective and hope you are benefited from my perspective and the perspectives of other people. I've listed all twenty techniques in chapter 14. And I hope this book is of great help to you. If not, then do not give up. There are other techniques out there to help with voices, and I would encourage you to find the safer ones, and hopefully have the help you need and want in dealing with voices. With or without voices, keep on living and breathing!

www.ingramcontent.com/pod-product-compliance
Lightning Source LLC
Chambersburg PA
CBHW021928170526
45157CB00005B/2232